On the Move

by Deborah Heiligman • illustrated by Lizzy Rockwell

For Benjamin,
who is always on the move
—D.H.

For Nicholas and Nigel
—L.R.

With thanks to Eliza Oursler
for her expert advice

The illustrations for this book were done in pencil and watercolor on T. H. Saunders watercolor paper.

The *Let's-Read-and-Find-Out Science* book series was originated by Dr. Franklyn M. Branley, Astronomer Emeritus and former Chairman of the American Museum–Hayden Planetarium, and was formerly co-edited by him and Dr. Roma Gans, Professor Emeritus of Childhood Education, Teachers College, Columbia University. Text and illustrations for each of the books in the series are checked for accuracy by an expert in the relevant field. For a complete catalog of Let's-Read-and-Find-Out Science books, write to HarperCollins Children's Books, 10 East 53rd Street, New York, NY 10022.

HarperCollins®, ♣®, and Let's Read-and-Find-Out Science® are trademarks of HarperCollins Publishers Inc.

On the Move

Library of Congress Cataloging-in-Publication Data
Heiligman, Deborah.
    On the move / by Deborah Heiligman ; illustrated by Lizzy Rockwell.
        p.   cm. — (Let's-read-and-find-out science. Stage 1)
    ISBN 0-06-024741-X. — ISBN 0-06-024742-8 (lib. bdg.)
    ISBN 0-06-445155-0 (pbk.)
    1. Human locomotion—Juvenile literature.   2. Animal locomotion—Juvenile literature.   [1. Human locomotion.   2. Animal locomotion.]   I. Rockwell, Lizzy, ill.   II. Title.   III. Series.
QP301.H35   1996   95-6738
612.7'6—dc20   CIP
                                                                                                                              AC

Typography by Elynn Cohen
1   2   3   4   5   6   7   8   9   10
❖
First Edition

On the
Move

I walk. I run. I ride my bike. I dance. I prance. I march.
I skip and hop and twirl around. I don't want to stop!
I am on the move.

I use my legs and my feet to move. I kick a ball.
I walk on my tiptoes. I twirl and whirl and dance.

6

I use my arms and hands. I throw a ball.
I catch it, too (sometimes).

I use my whole body
when I swing way up high—

8

and when I slide down fast.
 I move all day in many ways.
So do you. We are on the move!

When we were tiny babies,
we couldn't walk or run or even crawl.
We just waved our arms around, and
kicked our legs.

10

When we were
a few months old,
we rolled over.

COME ON,
YOU CAN DO IT!

A few months later,
we learned to sit up.

11

Babies usually crawl before they walk.
My mother says the first time I crawled,
I crawled right into the wall. *Ouch!* I never
did that again.

BOY, YOU'RE FAST!

My baby cousin crawled backward for a long time.
Now she is starting to walk. They call her a toddler
because she toddles. I think she wobbles.

Grandpa hobbles. He uses a cane to help him walk.
He walks slowly now. But he ran fast when he was a boy.

13

My friend can't walk at all. Her legs are weak.
She needs a wheelchair to move around. But her arms
are very strong. She can do a lot of pull-ups. She
shows me how.

14

I can't do pull-ups very well. But I practice.
The more I practice, the stronger I get.

I practice a forward roll, too.

I reach up
my arms,

put my chin
on my chest,

bend my knees,

tuck,   and roll.   I did it!

I jump, jump, jump in place.

I like to jump.
I jump up into the air.

18

I jump over a rope.

I jump down off a log.

19

I try to balance on the log, too.
Can I walk without falling off? It
helps to hold my arms out. I did it!

20

I jump off and I flap my
arms. I'm a bird! I feel
the air rushing past my
wings—I mean my arms.
I can't really fly, but it
feels good to pretend.

21

I pretend I'm a stork.
I stand on one foot, very still.
It's hard to stay that way.
You try it.

Now I hop.
I hop on one foot.

I hop on two.

I hop, hop, hop. I'm a kangaroo! Are you?

Frogs hop, too. But
they hop on all fours.

24

You have to crouch down to hop like a frog.
Ribbit! Ribbit!

Can you move like your favorite animal?

CLIP CLOP
CLIP CLOP

Can you gallop like a horse?

SSSSSSS

Can you wiggle like a worm and slither like a snake?

Can you waddle
like a duck?

Can you walk like a dog or cat?
You'll need four legs for that!

27

I swing like a monkey.
I hang upside down, too.
(I use my legs since I
don't have a tail.)

I climb like a monkey way up high, and back down again. I'm hungry for a banana!

But a burger looks good too.
I'm tired. I've been on the move all day.

31

After all this moving, my body
needs a rest. I stretch with my arms
and yawn. I crawl under the covers.
I will sleep well tonight!
Tomorrow I will be on the move again.